THE PIG

Artlist Collection

THE
PIG™

www.thepig-club.com

THIS IS A CARLTON BOOK

Published in 2009 by Carlton Books Limited
20 Mortimer Street
London W1T 3JW

10 9 8 7 6 5 4 3 2 1

The Pig Logo and Photographs © artist 2009
Text and design © Carlton Books Limited 2009

A CIP catalogue record for this book is available from
the British Library.

ISBN 978 1 84732 452 8

Publishing Manager: Penny Craig
Project Editor: Jennifer Barr
Managing Art Director: Lucy Coley
Design: Zoë Dissell
Production Controller: Luca Bazzoli

Printed in China

THE PIG

CARLTON
BOOKS

INTRODUCTION

Pigs are such loveable creatures. They are cute, playful, loyal and gregarious. Contrary to popular belief, they are not dirty, stupid or greedy but are actually among the cleanest and most intelligent animals on the planet. Indeed, pigs have been proved to be superior in intelligence to dogs, ranking behind only apes and dolphins. Pigs have even been trained to play computer games by using their snouts to move joysticks. They can also connect verbally with humans – if you give a pig a name and use it regularly, the pig will respond to it within a couple of weeks.

Pigs were among the first animals to be domesticated – as far back as 9,000BC – and have long been regarded as a symbol of happiness and luck. Since pigs are such good foragers with an excellent sense of smell, they are often employed to find truffles in Europe and North America. There are approximately 2 billion in the world – a total of 6.54 million pigs in the United States alone – pigs being found on every continent except Antarctica. Millions of these are kept as domestic pets – celebrity pig owners have included George Clooney, Jessica Simpson and Paris Hilton. Pigs are also hugely popular creatures on film and television, starring in the Oscar-winning movie *Babe* and its sequel *Babe: Pig in the City*.

The Pig comes from the same group who bought you *The Dog* and *The Cat*, two of the coolest and most successful licensed properties from Japan. Featuring more than 100 photographs taken with a fish-eye lens of pigs playing, being dressed up, relaxing and socializing, and including some fascinating facts about pigs in general and some information on traditional breeds too, *The Pig* will help you get to know these delightful animals as they really are – from snout to tail. The results are guaranteed to have you squealing with delight.

 It's all right for you two, but I really *am* afraid of the Big Bad Wolf.

THE ORIGINS OF THE PIG

Pigs are one of the oldest forms of livestock, domesticated earlier than cows. The Middle Eastern wild boar *Sus scrofa* is the ancestor of the various domesticated breeds, the farming of which was first practised in the Middle East and China at least 11,000 years ago. The pig went on to become a common farm animal in ancient Greece and Rome and through the centuries some 400 different breeds have been raised, although more than three-quarters of these have disappeared in recent decades owing to the development of intensive rearing systems.

I've had to be bottle-fed today.
It's Mum's day off.

We met on the internet, but we tell everyone we bumped into each other in the farmyard.

9

SUPER SNOUTS

Pigs are omnivores and spend much of their time rooting and foraging. Strengthened by a special bone and a disk of cartilage in the tip, the snout is a highly sensitive tool that enables them to find a wide variety of foods such as earthworms, mushrooms, grasses, leaves, roots, fruit, garbage, reptiles and rodents. They can learn where food is located simply by watching each other. Trials have shown that if one pig finds food, the others will "follow the leader" rather than search on their own. An average pig eats a ton of food every year but water remains the most important component of its diet, a pig's body comprising between one-half and two-thirds water.

He may be a lazy pig,
but he's *my* lazy pig.

Men can be such boars but a girl still has to make an effort.

Do you see what I've done with my ears? Clever, isn't it? Do you think I should enter a talent competition?

PORTLY PORKER

Pigs are distantly related to the hippopotamus. Their stocky, barrel-like bodies usually tip the scales at between 136–317 kilos (300–700 pounds) but can weigh considerably more. The largest pig on record was a Poland-China hog named "Big Bill", owned by Elias Buford Butler of Jackson, Tennessee. "Big Bill" stood 1.52 metres (five feet) tall from trotter to shoulder and measured 2.75 metres (nine feet) in length from tip of snout to tail. In 1933, he weighed in at a colossal 1,158 kilos (2,552 pounds) and was so large that he dragged his belly on the ground. He was scheduled to be exhibited at the Chicago World Fair but broke a leg and could not attend. Despite their heavy frames, pigs can reach speeds of 11 kilometres-per-hour (seven miles-per-hour).

I can stare you out if I really try.

This is *my* cushion, and I shall defend it to the last.

It's true – gentlemen do prefer blondes.

17

Please don't leave me behind.
I want to come, too.

NEST BUILDING

In the wild, a sow is very particular about the location and quality of her nest and may walk up to nine kilometres (six miles) before finding a sufficiently isolated and protected spot. She can then take as long as ten hours to build the nest. Once the piglets are born, mother and children remain in the nest for up to two weeks. Play is an important part of the piglets' lives – manifesting itself in fighting, chasing and frolicking – and the bond between them and their mother is very strong. Sometimes mother sows will even band together with other sows to create extended families. A sow that has never given birth is known as a gilt.

19

Did you know I sang
backing vocals for
Madonna?

Replica sports shirts are so expensive these days that Mum could only afford the hat and scarf.

21

No, she hasn't got a dirty nose.
It's her markings. Okay?

VOCAL REPERTOIRE

Although pigs are most commonly associated with an "oink oink" sound, they are actually extremely vocal creatures with a wide repertoire of "words". Their language includes grunts, squeals, roars, snarls and snorts, not to mention the occasional bout of jaw chomping and teeth clacking. Boars (male pigs) use distinctive mating noises to attract females who in turn use a special grunt to tell their piglets it is time to suckle. Piglets themselves have a special distress call they use when separated from their mother. And since a pig's squeal can reach up to 115 decibels – three decibels higher than Concorde – this is not to be sniffed at.

23

I knew it was a stupid idea to go to a fancy dress party as a zebra.

 Check if the coast is clear and then we'll raid the vegetable patch.

25

PIG OLYMPICS

Each year Russia hosts its very own Pig Olympics, featuring events such as pig racing, pig swimming and pigball – which is similar to football but has two teams of five piglets chasing a fish oil-covered ball with their snouts around a pen. The Sport-Pig Federation boasts around 100 members and draws competitors from as far away as South Africa. The good news about the Pig Olympics is that there are no losers: all the participants are used to breed a new generation of sporting pigs.

 I can hear every word you're saying about me, you know.

After a hard day at the trough you can't beat a nice little shuteye.

If I try to walk in this scarf, I'm worried that I'll trip over my own trotters.

29

What do you mean, you don't think it's my best side?

SOCIAL ACTIVITY

Pigs are party animals. They are very social creatures and it is thought that they can recognize and remember as many as 30 other pigs in their immediate vicinity. Pigs who know each other behave accordingly. Just as humans shake hands or hug, a pig may greet a friend by making nose-to-nose contact or by an affectionate display of grooming. They manage to establish a stable social group by evaluating each other's behaviour and by understanding which members of their herd are dominant and aggressive. Pigs are so communal that they even sleep huddled together.

This new nail varnish is
taking an eternity to dry.

I heard that if you blow in one ear, you can feel the draught out of the other.

OLD SPOTS

Thought to have originated in the nineteenth century, Gloucestershire Old Spots are one of the largest British pigs. Predominantly white in colour, their skin is marked with black spots, although in recent years selection has favoured less black, with only a few spots found on most of the pigs. Placid by nature, the breed also has a heavy drooped ear. Old Spots used to be called Orchard Pigs because they were partly raised on windfall apples. Indeed, local folklore suggests that the spots on their back were caused by falling fruits! The sows are known for large litters and high milk production.

 I'm playing the Kate Winslet role in our remake of *Titanic*.

35

I'm in love with the world's number one pig Elvis impersonator.

Don't you know that it's rude to look up a lady's nose?

DUROC PIGS

Found throughout the United States, Duroc pigs are a popular breed. This strain of pig was started in Saratoga County, New York, by Isaac Frink. He bought his first hogs in 1823 from Harry Kelsey, who at the time was keeping the famous thoroughbred stallion Duroc at his farm. When Frink took a shine to some of Kelsey's pigs and discovered they had no breed name, he called them Durocs in honour of the stallion. The breed shows considerable variation in colour, ranging from a light gold, almost yellow, to a very dark red. They also produce large litters.

Yes, I know: pearls before swine. Excuse me while I stop my sides from splitting.

Truffles. I can smell truffles.
Who's got truffles?

I may not have much weight to throw about yet, but I can do my best.

I've just got this sneaky feeling
that someone's watching me.

LARGEST LITTER

A pig's gestation period lasts 115 days, allowing a sow to have two or even three litters of piglets a year. Each litter consists of anything between six and 20 young, although the largest litter of piglets ever farrowed was an astonishing 37 by a sow on a farm in Australia. In total 36 piglets were born alive and 33 of them survived. Such is a pig's capacity for giving birth that a sow breeding continuously for ten years could, if all her daughters did likewise, have seven million descendants in that time!

43

It's all right for you, you're wide awake in the morning but I'm still half asleep.

 44

45

I happen to be a superior breed of pig.
I told my agent I'm too good for this book.

HEROIC PRISCILLA

Owned by Victoria Herberta of Houston, Texas, Priscilla the pig had always been fond of swimming, and she demonstrated her talent one day in 1983 at Lake Somerville after a boy stepped into a hole and slipped into the water. Hearing his screams, Priscilla swam towards him, allowing him to grab her harness and be pulled to safety. Her heroism saw her become the first pig to receive the William O. Fillman Award, given to pets that perform an act of life-saving courage, and the city of Houston declared a "Priscilla the Pig Day". Sadly, like so many celebrities, Priscilla developed a drug habit. She began devouring morning glory plants and had to be sent away to detoxify.

 Christmas always gives me
something to smile about.

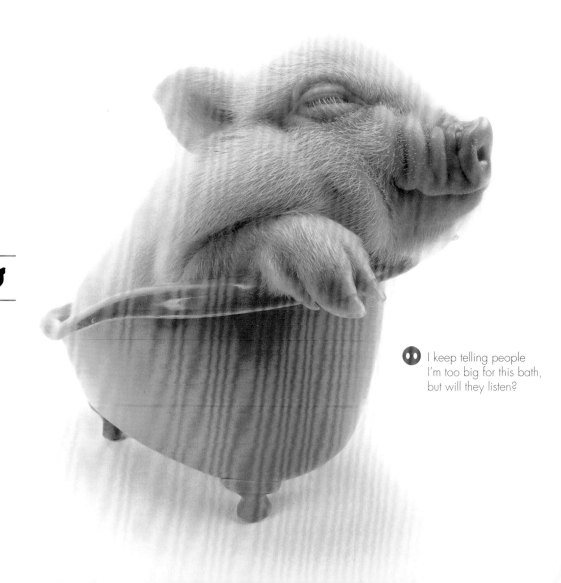

I keep telling people
I'm too big for this bath,
but will they listen?

Very well, I admit it. I was the one who cried "Wee! Wee! Wee!" all the way home.

PIGS FOR ALL SEASONS

In the sixteenth century, pigs were often employed in towns as efficient scavengers, acting as animal refuse services to clean up the neighbourhood. This practice continued in Ireland until the 1820s. Some pigs were trained to act like gundogs and the sturdier breeds were encouraged to pull carriages as an alternative to the horse. Performing pigs were not uncommon either. It was reported that a performing pig in London's Pall Mall in the 1790s attracted sizeable and profitable audiences, while another pig was renowned for its love of herrings and would perform a variety of tricks in order to receive its favourite reward.

Sometimes you just have to take the
weight off your feet – particularly
when you've got four of them.

What a lousy place to have a birthmark!

 Mirror, mirror on the wall,
who is the fairest piggy of them all?

CLOONEY'S PET

For 18 years, actor George Clooney shared his Hollywood Hills home with a Vietnamese potbellied pig called Max. Bought by Clooney in 1988 as a gift for his former girlfriend Kelly Preston, the 136-kilo (300-pound) animal had a special cattle-pen and his own corner in the garage but would sometimes climb into Clooney's bed. "Max the star", as Clooney often referred to him, was even credited with saving the actor's life by waking him during a 1994 earthquake that hit Los Angeles. A few months before his death in 2006, Max was taken for a flight in John Travolta's private jet, and according to Clooney, he "absolutely loved it."

If we just close our eyes and wish
hard enough...

Keep over your own side of the bath
or I'll give you the tap end next time.

 You don't think my fringe
is too long, do you?

57

I don't care what you say,
I think I scrub up pretty well.

PIGS AND SUNBURN

Pigs, walruses and light-coloured horses are the only animals other than humans that can suffer from sunburn. Although pale-skinned pigs are prone to sunburn, pigs don't actually sweat at all – for the simple reason that they have no sweat glands. Instead, to cool off they wallow in mud baths, which leads to their unjustified reputation as dirty animals. But if they have somewhere to shelter from the sun, they are extremely hygienic, even refusing to excrete their waste near their living or eating areas. In fact, give a pig a nice field full of clover and it would be like… well, a pig in clover.

Yes, I know I'm standing on tiptoe. I'm trying to look tall and mean.

I'm hoping that not even the Big Bad Wolf would want to make an old lady homeless.

That's a bit of luck –
this basket is just my size.

BOISTEROUS TAMWORTHS

Tamworth pigs originated in Ireland where they were called "Irish Grazers". Around 1812, Sir Robert Peel, impressed with their characteristics, imported some of them and started to breed them on his estate at Tamworth, Warwickshire. In appearance the Tamworth has a red coat and a long, straight snout. The sows are excellent mothers and do a good job of suckling their litters. Regarded by many as the aristocrat of the pig world, the Tamworth is a boisterous pig whose popularity has fluctuated through the years to the extent that it has sometimes been close to extinction in the UK. However, it remains highly valued in the United States.

It always looks so easy
when they do it in Westerns.

I never wanted to be a singer.
I really wanted to be a dancer,
but they said I had two left feet…
which is true.

TUSK GROWTH

Pigs have a full set of 44 teeth. In males, the canine teeth, or tusks, grow continually and are sharpened by the uppers and lowers rubbing against each other. The tusks, which are used for protection and for digging, appear as protrusions under the sides of the lip at around a year old. The speed of tusk growth varies according to the breed but by the time the animal is four or five years old, the tusks will be fairly large. Some pig breeders choose to remove the tusks of even the most gentle boar to prevent accidental damage to handlers or other pigs, and this is usually carried out using razor wire.

I painted it myself.
Do you like it?

67

Just give me five more minutes and then I promise I'll get up.

68

I went up to this baby goat the other day and said, "Here's looking at you, kid." He didn't think it was very funny.

Mmmm! I'm dreaming about rotting leaves, mouldy apples and potato peelings.

70

BRITISH SADDLEBACK

Created by an amalgamation of the Essex and Wessex breeds in 1967, the British Saddleback is distinguishable by the white belt around its body. The width of the belt can vary considerably from a thin strip to one covering almost the entire body. Lop-eared, hardy, docile and good grazers, Saddlebacks are ideal pigs for cold climates. They also have a fine reputation as mothers, giving birth to large litters. Despite its virtues, the breed is considered to be "at risk" as there are fewer than 500 breeding British Saddleback sows in the UK.

71

 It took a whole day's shopping to find an outfit that matched my snout.

I don't suppose you know where I can buy a jar of anti-wrinkle cream?

73

UNLIKELY PLAYMATES

At a zoo in Bangkok, a fully grown tigress recently formed an unlikely friendship with a piglet. To ensure that the pig came to no harm, keepers dressed it in a tiger-print jacket to convince the big cat that it was one of her cubs! Meanwhile at a zoo in Guangzhou, China, a mother pig returned the compliment by acting as a surrogate mum to three tiger cubs that had been rejected by their own mother. As well as drinking the sow's milk, the cubs also enjoyed playing with their piglet foster brothers and sisters.

If I'm young, why have I got such a hairy nose?

And now I'd like to show you some moves from my latest starring role in *Swine Lake*.

 I can smell food...

WEDDED BLISS

To mark the Year of the Pig in 2007, two pot-bellied pigs were married at a special ceremony in Taiwan, presided over by the parish priest and watched by more than 100 guests. Farm owner Shu Wen-chun had been scouring the internet for months to look for a suitable bride for his male pig Shui Fu-ko, and after sorting through dozens of potential suitors, he settled on Huang Pu-pu, a 10-kilo (23-pound) porker from north-west Taiwan. After the ceremony, Shu said: "They will be living in a luxurious place with a pond in the front and a mountain at the back and they will have a happy life."

Forget it, cop, I ain't squealing.

I thought skateboards were supposed to have wheels. They've sold me one without wheels!

 Who could be lucky enough to find me in their Christmas stocking?

A PIG CALLED BABE

The 1995 Australian movie *Babe* – the story of a little pig that wanted to be a sheepdog – was a huge critical and box office success. Based on the book by Dick King-Smith, it grossed $254,134,910 worldwide and won an Oscar. It led to a 1998 sequel, *Babe: Pig in the City*, and both films were nominated for awards in the UK, the United States and Australia. Filming *Babe* wasn't always easy, partly because piglets grow so quickly, sometimes multiplying their birth weight four times in the first two weeks. So it was impossible to use just one "Babe" over the long filming schedule – and in fact a total of no fewer than 48 little pigs took turns at playing the title role.

 Well, I think I look like Cher.

83

If the Queen of England is ever ill, I can step in at a moment's notice.

Or do you think the blue is more my colour?

85

PIG SQUEALING CHAMPIONSHIPS

Every year in the town of Trie-sur-Baïse in southwestern France, humans don pig costumes to take part in the Pig Squealing Championships. Competitors from all over the country stand before a microphone and attempt to mimic the various sounds of a pig – including squeals, grunts and snuffles. The winner is the one deemed by the panel of judges to have best captured the cries associated with the different stages of a pig's life.

I told them Pinkbeard didn't sound menacing enough for a pirate.

Is there a problem with me having a woolly duck as a friend?

89

Can't you see I'm bristling
with indignation?

POT BELLIES

Vietnamese pot-bellied pigs have become popular pets over the last 30 years. With upright ears, a short snout and a straight tail that wags with pleasure, they are the size of a medium dog but their bodies are considerably denser, weighing anything up to 136 kilos (300 pounds). Although they naturally have pot bellies, these should not be so large as to touch the ground. A pig with rolls of fat over its eyes is another indicator of excess weight – the eye sockets of a pot-bellied pig should be visible at all times. Vietnamese pot-bellied pigs can be taught to sit and fetch, they can be trained to use litter boxes, and many owners take them for walks with harnesses and dog leads but, unlike small dogs, they do not like to be picked up and held.

Who'd have thought it? Pigs *can* fly.

Do you like my new suit? I think it makes me look quite distinguished.

Yes, I suppose
I can be pig-headed.

I was crowned Miss Twirled – until they saw my tail was actually straight.

LANDRACE PIGS

The Landrace was first imported into Britain from Scandinavia in 1949. It is easily recognizable by its long straight snout, semi-lop ears and very fine white hair, which unfortunately does little to protect it during cold weather. Equally, its pink skin makes it vulnerable to sunburn. Some Landrace bloodlines also carry the genetics for Porcine Stress Syndrome, a condition which causes the pig's muscles to overheat when the animal is under stress or subjected to strenuous activity. Consequently the Landrace is not considered suitable for free-range farming.

95

 You don't think this is a bit over-the-top just for a trip to the trough?

It was so kind of Miss Piggy to give me a few of her old clothes.

No, I think you'll find
that I was first in the queue.

TINY TROTTERS

The smallest breed of pig is the Mini Maialino, which weighs only 9 kilos (20 pounds) when fully grown. But the world's most popular miniature pig is the Kune Kune from New Zealand. Coming in a variety of colours from cream to black, sandy to spotted, these docile little pigs have a maximum weight of about 11 kilos (25 pounds) and grow to a height of just 60 centimetres (two feet). Kune Kunes love human company and are very easy to train. A distinguishing feature in many of the breed is a pair of tassels (called "piri-piris") under their chin, rather like those of a goat.

Now what's my personal stylist's phone number?

100

Please, just a little kiss.

 See, it's not only that Jim Carrey who can pull funny faces.

GREEN PIGS

In 2005, scientists in Taiwan bred three fluorescent green pigs by planting genetic material from jellyfish into ordinary pig embryos. In daylight, the pigs' eyes, teeth and trotters look green, and their skin has a greenish tinge, but in the dark under a blue light their whole bodies glow brightly. Even the pigs' internal organs are green. The scientists hope the glowing pigs will help further the study of stem cell research.

103

 Aloha, welcome to Hawaii.

 Nobody will tell me why my legs are a different colour to the rest of my body.

105

 Come on, let's hold trotters.

DESIGNER SWINE

A 2008 art exhibition in China featured a group of pigs tattooed with Louis Vuitton logos and other designs. "Art Farm" was the work of China-based Belgian artist Wim Delvoye who tattooed eight pigs when they were still small. Delvoye and his team then watched his designs grow along with the animals. He insisted that the pigs were treated humanely and were given sedatives before being tattooed. However plans to transfer the exhibit to Shanghai were scrapped after gallery owners there received a string of complaints from the public.

It's my chance to hog the limelight.

 It's tough when you fall on hard times
and have to sleep rough.

WOOLLY MANGALITZAS

At first glance, the Austro-Hungarian Mangalitza looks more like a sheep than a pig. The blonde variety of the breed has a mass of golden curls covering its entire body, making it a natural successor to the Lincolnshire Curly Coat which became extinct in 1972. A large, hardy pig, the Mangalitza was once in huge demand all over Europe and even traded on the Vienna Stock Exchange. At its peak, 100,000 animals a year were sold from Hungary to the west. Mangalitzas come in three different colours – Swallow Bellied (black with a white underbelly), Blonde (a dense coat varying from grey to yellow) and Red (with a thick reddish coat).

I'm beginning to wish I'd never bought that workout DVD.

What do you call a pig that's lost its voice? – Disgruntled. Geddit?

Here's a better one. Why is getting up at five o'clock in the morning like a pig's tail? – It's twirly.

SOOTHING SOUNDS

Some farmers have discovered that playing music to pigs while they are eating keeps them happy. Nguyen Chi Cong, who has run a farm near Ho Chi Minh City in Vietnam for over 20 years, began playing classical works by Beethoven, Mozart and Schubert over loudspeakers to his 3,000 pigs for six hours a day and found that it had a soothing effect on the animals. "I saw that my pigs started eating more and that they were gaining weight faster than usual," he said, adding that other farmers in the area have flocked to his premises to learn about the technique.

This is a good place to hide. I don't think anyone can see me here, do you?

115

Has anyone ever told you that you're a fantastic kisser?

I can get around the supermarket much quicker in one of these.

117

ROYAL APPROVAL

Said to have been discovered by the army of English general Oliver Cromwell in their winter quarters at Reading back in the seventeenth century, the Berkshire is a black pig with impressive pricked up ears, a white blaze on its face, a white-tipped tail and white feet. Queen Victoria bred a Berkshire boar called Ace of Spades, and for many years the breed was a favourite of the British Royal Family who kept a large herd of Berkshires at Windsor Castle. The Berkshire is renowned as a pig of great personality, and it was probably this which inspired author George Orwell to make the character Napoleon in his novel *Animal Farm* a Berkshire boar.

118

You have to look your best on your wedding day.

Who's stolen my piano?

I'm all dressed up to appear on that new reality TV show, *Pig Brother*.

 He thinks I haven't seen him eyeing up that Gloucestershire Old Spot in the next pen.

122

THE TAMWORTH TWO

While being unloaded from a truck in Malmesbury, Wiltshire, in 1998, a pair of Tamworth pigs made a daring escape and went on the run for a week, sparking huge media interest across Britain. The pair – who were subsequently given the names of Butch and Sundance – were five-month-old sister and brother Tamworths. The Tamworth Two (as they became known) escaped by squeezing through a fence, then swimming across the River Avon and running through gardens. After hiding out in a dense thicket, they were eventually recaptured a week later while foraging in the garden of a local couple. By then their fame was such that they were sent to live in an animal sanctuary.

Does anyone want a
game of pigball?

125

Wake me when it's feeding time.

ALMOST HUMAN

Pigs can live for up to 15 years and as well as having an excellent sense of smell, they possess good hearing and eyesight. Pigs and humans have much in common. Like humans, pigs have single-chambered stomachs, their body temperature is similar to ours and many other anatomical and physiological features of the pig are identical to those of humans. Also, pigs investigate with their mouths, just like human babies – anything unusual that is encountered is likely to be tasted or nibbled. Unfortunately pigs are also susceptible to some diseases which affect humans, including pneumonia and bronchitis.

 It may be small, but it's home.

I told you before:
no make-up artist,
no photo.

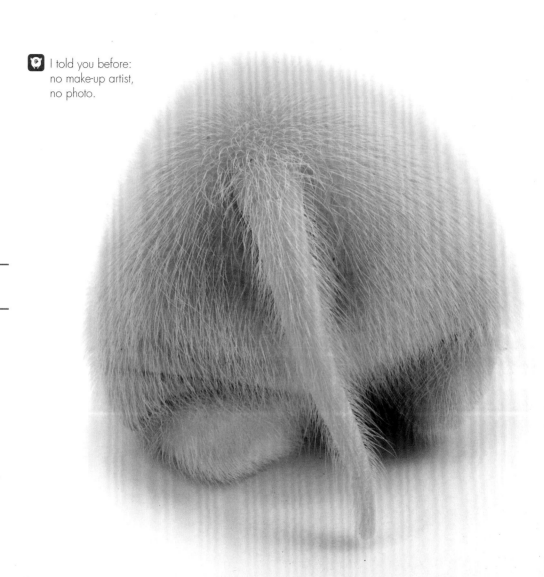